UNDERSTANDING MPOX:
A COMPREHENSIVE GUIDE TO THE INNOVATIONS IN PUBLIC HEALTH

BY

VINCENT MOREAU

COPYRIGHT

All rights reserved. No part of this publication may be reproduced, distributed, or transmitted in any form or by any means, including photocopying, recording, or other electronic or mechanical methods, without the prior written permission of the publisher, except in the case of brief quotations embodied in critical reviews and certain other noncommercial uses permitted by copyright law. Copyright © Vincent Moreau 2024.

FOREWORD

In an era where global health challenges are becoming increasingly complex and interconnected, understanding diseases like mpox (formerly known as monkeypox) is more crucial than ever. This book aims to provide a comprehensive overview of mpox, its transmission, symptoms, and the public health strategies necessary to combat its spread.

As we navigate through the intricate web of zoonotic diseases, it is essential to recognize that our health is intertwined with the health of our environment and the animal kingdom. The resurgence of mpox serves as a reminder of the delicate balance we must maintain in our ecosystems and the importance of vigilance in public health.

The content within these pages is designed not only to educate but also to empower readers—whether they are healthcare professionals, policymakers, or concerned citizens. With a clear glossary of terms, contact information for vital health organizations, and practical insights into prevention and response, this book serves as a valuable resource for anyone seeking to deepen their understanding of mpox and its implications.

In addition, this work highlights the importance of community engagement and the role of mental health in times of crisis. As we confront the realities of infectious diseases, it is imperative that we foster a supportive environment for those affected and ensure that mental well-being is prioritized alongside physical health.

I invite you to explore the chapters ahead with an open mind and a commitment to learning. Together, we can build a more informed and resilient society capable of facing the challenges posed by emerging infectious diseases like mpox.

With gratitude for your interest in this vital topic,

Vincent Moreau

ACKNOWLEDGEMENTS

I would like to express my heartfelt gratitude to everyone who contributed to the creation of this book.

First and foremost, I extend my thanks to the healthcare professionals, researchers, and public health officials whose tireless work and dedication to understanding and combating mpox have inspired this project. Your insights and expertise have been invaluable in shaping the content presented here.

A special thank you goes to the organizations and institutions that provided essential data, resources, and support. Your commitment to advancing public health knowledge has made this work possible.

I am also grateful to my colleagues and mentors for their guidance and encouragement throughout this journey. Your feedback and collaboration have enriched this book in countless ways.

To the individuals and communities affected by mpox, your resilience and stories have served as a powerful reminder of the human experience behind the statistics. This book is dedicated to you, with the hope that it will contribute to greater awareness and understanding.

Lastly, I would like to thank my family and friends for their unwavering support and patience during the writing process. Your belief in this project has kept me motivated.

Together, we can foster a deeper understanding of mpox and work towards a healthier future for all.

With sincere appreciation,

Vincent Moreau

TABLE OF CONTENTS

COPYRIGHT 2
FOREWORD 3
ACKNOWLEDGEMENTS 5
INTRODUCTION TO MPOX 7
CHAPTER 2 11
CHAPTER 3 16
CHAPTER 4 20
CHAPTER 5 25
CHAPTER 6 31
CHAPTER 7 37
CHAPTER 8 42
CHAPTER 9 46
CHAPTER 10 50

INTRODUCTION TO MPOX

OVERVIEW OF MPOX

Mpox, previously known as monkeypox, is a viral zoonotic disease caused by the pox virus, a member of the Orthopoxvirus genus. Characterized by symptoms that can range from mild to severe, mpox manifests primarily through fever, rash, and swollen lymph nodes, often resembling the clinical presentation of smallpox. While it was traditionally confined to certain regions in Africa, recent outbreaks have raised concerns about its global reach and implications for public health.

The virus is transmitted to humans through direct contact with infected animals, typically rodents or primates, but human-to-human transmission is also possible through respiratory droplets, bodily fluids, and skin lesions. The emergence of mpox as a public health concern underscores the importance of understanding zoonotic diseases and their potential to cross species barriers.

Historical Context

The history of mpox dates back to its first identification in laboratory monkeys in 1958, which led to its initial name. The first documented human case occurred in 1970 in the Democratic Republic of the Congo. Since then, sporadic cases have been reported primarily in Central and West Africa, often linked to animal reservoirs or human contact with infected individuals.

For decades, mpox remained relatively obscure outside its endemic regions. However, the landscape began to shift dramatically in the early 21st century. In 2003, an outbreak in the United States traced back to imported animals highlighted the virus's potential for international spread. This incident marked a pivotal moment in public awareness and surveillance efforts surrounding mpox.

As globalization and environmental changes continue to reshape our interactions with wildlife, the risk of zoonotic diseases like mpox has become more pronounced. The recent outbreaks in non-endemic countries have sparked renewed interest and urgency in studying this virus, prompting health authorities to strengthen surveillance systems and develop effective preventive measures.

In this book, we will delve deeper into the complexities of mpox—its virology, clinical presentation, public health implications, and the multifaceted strategies required to combat its spread. Understanding mpox is not only critical for managing current outbreaks but also for preparing for future challenges posed by emerging infectious diseases.

CHAPTER 2

VIROLOGY AND TRANSMISSION

The MPox Virus: Structure and Characteristics

The mpox virus is a double-stranded DNA virus belonging to the Orthopoxvirus genus, which also includes variola (the smallpox virus), vaccinia (used in the smallpox vaccine), and cowpox viruses. Here are some key characteristics:

-Structure: The pox virus has a complex structure, characterized by an oval or brick-like shape. It possesses a lipid envelope that contains surface proteins crucial for binding to host cells. The viral genome encodes various proteins that facilitate replication and immune evasion.

-Genomic Features: The genome of the mpox virus is approximately 197 kb in length and contains genes responsible for viral replication, immune modulation, and virulence. Genetic studies have shown that there are two main clades: the Central African clade and the West African clade, with differences in pathogenicity and transmission dynamics.

-Stability: The mpox virus is relatively stable in the environment, which aids in its transmission. It can survive on surfaces for extended periods, making it important to consider environmental factors in infection control.

Modes of Transmission

Mpox can be transmitted through several routes:

1. Animal-to-Human Transmission:
    - Direct Contact: Handling infected animals (e.g., rodents or primates) can lead to transmission through bites, scratches, or contact with bodily fluids.
    - Consumption of Infected Meat: Eating undercooked meat from infected animals poses a risk.

2. Human-to-Human Transmission:
    - Respiratory Droplets: Close contact with an infected person can result in transmission through respiratory droplets, especially during prolonged face-to-face interactions.
    - Skin Lesions and Bodily Fluids: Direct contact with skin lesions or bodily fluids of an infected person can facilitate transmission.
    - Fomites: Contaminated objects or surfaces can harbor the virus, contributing to indirect transmission.

3. Vertical Transmission: There is potential for transmission from an infected mother to her fetus during pregnancy or childbirth.

Risk Factors for Infection

Certain factors increase the likelihood of mpox infection:

- Geographic Location: Individuals living in or traveling to endemic regions (Central and West Africa) are at higher risk.

- Occupational Exposure: Healthcare workers, laboratory personnel, and those involved in wildlife handling or trade are particularly vulnerable due to their increased exposure to infected individuals or animals.

- Close Contact: Engaging in close physical contact with infected individuals—such as caregivers or family members—heightens the risk.

- Immunocompromised Status: Individuals with weakened immune systems may be more susceptible to severe disease upon infection.

- Cultural Practices: Certain cultural practices involving animal handling or consumption may elevate the risk of zoonotic transmission.

Understanding these aspects of mpox virology and transmission is crucial for developing effective prevention strategies and public health interventions to control its spread.

CHAPTER 3

SYMPTOMS AND DIAGNOSIS OF MPOX

Clinical Manifestations

The clinical presentation of mpox can vary, but common symptoms typically appear 5 to 21 days after exposure to the virus. The disease often progresses through several stages:

1. Prodromal Phase (1-5 days):
    - Fever
    - Chills
    - Headache
    - Muscle aches
    - Fatigue
    - Lymphadenopathy (swollen lymph nodes)

2. Rash Phase:
    - A characteristic rash usually develops within 1-3 days after fever onset.

- The rash typically starts as flat lesions (macules) that progress to raised bumps (papules), then vesicles (blisters), and finally pustules before crusting over.
- Lesions can appear on the face, palms, soles, and other parts of the body.

3. Resolution:
- The rash eventually scabs over and falls off, typically leaving no scars.
- Full recovery can take 2-4 weeks.

Diagnostic Methods

Diagnosis of mpox is primarily based on clinical evaluation and laboratory testing:

1. Clinical Assessment:
- A thorough history of exposure (travel to endemic areas, contact with infected individuals or animals) and symptom evaluation.

2. Laboratory Testing:
   - PCR (Polymerase Chain Reaction): The gold standard for diagnosis, PCR tests detect viral DNA from skin lesions, blood, or other bodily fluids.
   - Serology: Antibody tests can help identify previous infections but are not typically used for acute diagnosis.
   - Virus Isolation: Culturing the virus from lesion samples can confirm the diagnosis, though it is less commonly performed due to biosafety concerns.

3. Imaging Studies:
   - While not routinely used for diagnosis, imaging may assist in assessing complications or associated conditions.

Differential Diagnosis

Several conditions can mimic the symptoms of mpox, making differential diagnosis essential:

1. Other Poxviruses:
   - Variola virus (smallpox)
   - Vaccinia virus (from vaccination)

2. Viral Infections:
- Chickenpox (varicella)
- Herpes simplex virus (HSV)
- Coxsackievirus infections

3. Bacterial Infections:
- Impetigo
- Folliculitis

4. Other Dermatoses:
- Scabies
- Contact dermatitis
- Drug eruptions

5. Zoonotic Infections:
- Rabies (in cases of animal bites)
- Tularemia

Accurate diagnosis is crucial for appropriate management and public health response, especially given the potential for outbreaks and zoonotic transmission.

CHAPTER 4

PUBLIC HEALTH IMPACT OF MPOX

Epidemiology of Mpox

Mpox (formerly known as monkeypox) is caused by the pox virus, which belongs to the Orthopoxvirus genus. The disease was first identified in laboratory monkeys in 1958 but is primarily associated with rodents in Africa. Key epidemiological points include:

- Transmission: Mpox can be transmitted to humans through direct contact with infected animals (e.g., rodents), human-to-human contact, or contaminated materials. Respiratory droplets can facilitate transmission during prolonged face-to-face interactions.
- Geographic Distribution: Historically, mpox has been endemic to central and west African countries, including the Democratic Republic of the Congo and Nigeria. However, cases have

been reported outside these regions due to international travel and trade.
- Reservoirs: While rodents are considered the primary reservoirs, other animals, such as primates, may also play a role in transmission.

Recent Outbreaks and Trends

Recent outbreaks have highlighted the changing landscape of mpox epidemiology:

- 2022 Global Outbreak: In mid-2022, a significant outbreak occurred outside endemic regions, affecting multiple countries across Europe, North America, and beyond. The outbreak was characterized by:
- Rapid spread among men who have sex with men (MSM), leading to increased awareness and concern.
- A shift from traditional epidemiological patterns, with cases reported in non-endemic countries.

- Case Numbers: Thousands of cases were reported globally during the 2022 outbreak, prompting public health responses and vaccination campaigns in affected areas.

- Trends: The outbreak underscored the importance of surveillance and prompt identification of cases to prevent further transmission. It also highlighted the need for targeted public health messaging, particularly within high-risk populations.

Global Response and Preparedness

The global response to mpox has evolved significantly in light of recent outbreaks:

1. Surveillance and Reporting:
    - Enhanced surveillance systems were implemented to track cases and identify potential outbreaks rapidly.
    - Countries were encouraged to report suspected cases to facilitate international monitoring.

2. Vaccination Strategies:
    - Vaccines previously used for smallpox (e.g., JYNNEOS) were recommended for high-risk populations to prevent mpox infection.
    - Vaccination campaigns were launched in affected regions, targeting MSM and healthcare workers.

3. Public Health Messaging:
    - Clear communication strategies were developed to inform the public about the risks of mpox, preventive measures, and available resources.
    - Educational campaigns aimed at reducing stigma associated with the disease were prioritized.

4. International Collaboration:
    - Organizations such as the World Health Organization (WHO) coordinated efforts among nations to share information, resources, and best practices.

- Research initiatives focused on understanding the virus's transmission dynamics and developing effective treatments and vaccines.

5. Preparedness Plans:
    - Countries updated their pandemic preparedness plans to include mpox response strategies.
    - Stockpiling of vaccines and antiviral treatments was emphasized to ensure rapid response capabilities in future outbreaks.

The public health impact of mpox is significant, particularly given its potential for outbreaks beyond endemic regions. Continued vigilance, research, and international collaboration are essential to mitigate risks and respond effectively to future challenges posed by this zoonotic disease.

CHAPTER 5

PREVENTION AND CONTROL MEASURES FOR MPOX

Vaccination Strategies

1. Vaccination Availability:
   - Vaccines such as JYNNEOS (MVA-BN) and ACAM2000, which were originally developed for smallpox, are effective against mpox.
   - Vaccination programs are prioritized for high-risk populations, including healthcare workers and individuals with potential exposure.

2. Post-Exposure Prophylaxis (PEP):
   - Individuals who have been exposed to mpox may receive vaccination within 14 days of exposure to reduce the risk of developing the disease.

- PEP is particularly important in outbreak settings to contain the spread.

3. Pre-Exposure Prophylaxis (PrEP):
    - Vaccination is recommended for individuals at higher risk of exposure, such as those in close contact with infected individuals or those engaging in high-risk behaviors.

4. Public Health Campaigns:
    - Awareness campaigns to promote vaccination among vulnerable groups, emphasizing the safety and efficacy of vaccines.

Public Health Guidelines

1. Surveillance and Reporting:
    - Enhanced surveillance systems to monitor mpox cases and report them promptly to health authorities.
    - Regular updates on case numbers and outbreak status to inform public health responses.

2. Infection Control Measures:
    - Implementation of standard precautions in healthcare settings, including the use of personal protective equipment (PPE) when caring for suspected or confirmed cases.
    - Isolation of infected individuals to prevent transmission to others.

3. Travel Advisories:
    - Issuance of travel advisories for regions experiencing outbreaks, advising travelers on preventive measures and potential risks.

4. Guidelines for High-Risk Settings:
    - Specific guidelines for settings such as healthcare facilities, shelters, and correctional institutions to minimize transmission risk.

Community Awareness and Education

1. Public Awareness Campaigns:
   - Development of informational materials (brochures, posters, social media content) that explain what mpox is, its symptoms, transmission routes, and preventive measures.
   - Targeted messaging for communities at higher risk, particularly MSM and other vulnerable populations.

2. Workshops and Training:
   - Conducting workshops for community leaders, healthcare providers, and educators to disseminate information about mpox prevention and control.
   - Training sessions on recognizing symptoms and understanding when to seek medical care.

3. Reducing Stigma:
   - Initiatives aimed at reducing stigma associated with mpox, especially within affected communities, to encourage

individuals to seek care without fear of discrimination.
- Promoting inclusive messaging that emphasizes community solidarity and support.

4. Engagement with Community Organizations:
   - Collaboration with local organizations and advocacy groups to reach diverse populations effectively.
   - Empowering communities to take an active role in prevention efforts through grassroots initiatives.

5. Utilizing Technology:
   - Leveraging digital platforms (webinars, apps) to disseminate information quickly and engage younger audiences.
   - Creating online resources that provide updates on mpox and guidance on preventive measures.

A comprehensive approach combining vaccination strategies, public health guidelines, and community awareness is essential for preventing and controlling mpox outbreaks. By fostering collaboration among health authorities, communities, and individuals, the impact of mpox can be significantly reduced, ultimately protecting public health.

CHAPTER 6

TREATMENT OPTIONS FOR MPOX

Current Therapeutic Approaches

1. Antiviral Medications:
   - Tecovirimat (TPOXX): Approved for the treatment of orthopoxvirus infections, including mpox. It can reduce the duration of symptoms and is particularly beneficial for severe cases or those at high risk of complications.
   - Cidofovir: An antiviral that has shown effectiveness against various viral infections, including some poxviruses. It may be used in specific cases, especially when other treatments are not available.
   - Brincidofovir: An oral formulation of cidofovir that may be considered for treatment, especially in patients who cannot tolerate intravenous administration.

2. Supportive Care:
- Symptomatic treatment is crucial, including pain management, hydration, and treatment of secondary bacterial infections.
- Patients with severe skin lesions may require wound care and management of secondary infections.

3. Vaccination Post-Infection:
- Vaccination with JYNNEOS or ACAM2000 may be considered in certain cases to boost immune response, particularly in immunocompromised patients.

Management of Complications

1. Skin Lesions:
- Careful monitoring and management of skin lesions to prevent secondary infections.
- Use of topical antibiotics if bacterial infection is suspected.

2. Respiratory Complications:
   - Monitoring for respiratory distress, especially in cases with extensive lesions affecting the respiratory tract.
   - Supportive measures, including oxygen therapy and mechanical ventilation if necessary.

3. Neurological Complications:
   - Neurological symptoms should be promptly evaluated, and supportive care initiated as needed.
   - Consultation with neurology specialists may be warranted for severe cases.

4. Psychological Support:
   - Addressing mental health needs due to the stigma and anxiety associated with mpox infection.
   - Providing counseling services and support groups for affected individuals.

Future Directions in Treatment

1. Research on Novel Antivirals:
   - Continued research into new antiviral agents specifically targeting orthopoxviruses to enhance treatment options.
   - Exploration of combination therapies that could improve efficacy and reduce viral load more rapidly.

2. Immunotherapy:
   - Investigating the potential of immunotherapeutic approaches, including monoclonal antibodies, to enhance immune response against mpox.
   - Development of targeted therapies that could modulate the immune system's response to viral infections.

3. Clinical Trials:
   - Encouragement of clinical trials to evaluate the safety and efficacy of existing medications and new treatments in diverse populations.

- Gathering data on long-term outcomes and complications associated with mpox to inform future treatment guidelines.

4. Global Collaboration:
    - Enhancing international collaboration to share data, resources, and best practices for mpox management.
    - Establishing global surveillance systems to monitor mpox outbreaks and treatment efficacy.

5. Vaccine Development:
    - Research into next-generation vaccines that could provide broader protection or require fewer doses.
    - Exploring the use of vaccine platforms that could quickly adapt to emerging strains or related viruses.

Current treatment options for mpox focus on antiviral medications, supportive care, and management of complications. Ongoing research and innovation are essential to improve therapeutic approaches and outcomes for affected individuals. A collaborative effort among healthcare professionals, researchers, and public health authorities will drive advancements in mpox treatment and prevention.

CHAPTER 7

LIVING WITH MPOX

Coping with Infection

1. Understanding the Disease:
   - Educate yourself about mpox, including its symptoms, transmission, and treatment options. Knowledge can empower you and reduce anxiety.
   - Keep track of your symptoms and communicate openly with healthcare providers about any changes.

2. Managing Symptoms:
   - Follow medical advice for symptom management, including pain relief and wound care.
   - Maintain hydration and nutrition to support your recovery.

3. Self-Care Practices:
   - Engage in self-care activities that promote relaxation and well-being, such as gentle exercise, mindfulness, or meditation.
   - Prioritize rest to help your body heal.

4. Establishing a Routine:
   - Create a daily routine that includes time for self-care, work, and leisure activities to maintain a sense of normalcy.

Support Systems and Resources

1. Healthcare Support:
   - Regular check-ins with healthcare providers for monitoring and managing your condition.
   - Access to specialists if complications arise, such as dermatologists for skin lesions.

2. Support Groups:
   - Join support groups for individuals affected by mpox or similar conditions. Sharing experiences can provide emotional relief and practical advice.

- Online forums can also be a valuable resource for connecting with others in similar situations.

3. Mental Health Resources:
   - Consider speaking with a mental health professional who understands the challenges faced by those living with infectious diseases.
   - Look for counseling services that focus on coping strategies and emotional support.

4. Community Resources:
   - Local health departments may offer resources and information about mpox.
   - Nonprofit organizations may provide educational materials, support services, and advocacy.

Stigma and Mental Health Considerations

1. Addressing Stigma:
   - Understand that stigma surrounding mpox can stem from misinformation. Educating

others can help reduce negative perceptions.
- Share your story if comfortable, as personal narratives can challenge stereotypes and promote understanding.

2. Mental Health Impact:
   - Acknowledge the emotional toll of living with mpox, including feelings of isolation, anxiety, or depression.
   - Seek professional help if you experience persistent feelings of sadness or anxiety.

3. Building Resilience:
   - Practice self-compassion and recognize that it's normal to feel overwhelmed.
   - Engage in activities that boost your mood and foster connections with supportive friends or family members.

4. Mindfulness and Stress Reduction:
   - Incorporate mindfulness practices such as meditation, deep breathing exercises, or yoga to manage stress.

- Journaling can also be a helpful outlet for expressing feelings and processing experiences.

Living with mpox involves navigating physical, emotional, and social challenges. By utilizing available resources, fostering support systems, and addressing stigma, individuals can cope more effectively with the infection. Prioritizing mental health and self-care is essential for overall well-being during this time.

CHAPTER 8

CONCLUSION

Summary of Key Points

1. Understanding Mpox: Knowledge about mpox, including its symptoms, transmission, and treatment, is crucial for effective management and reducing anxiety.

2. Symptom Management: Following medical advice for symptom relief, maintaining hydration and nutrition, and practicing self-care can significantly aid recovery.

3. Support Systems: Accessing healthcare support, joining support groups, and utilizing mental health resources can provide emotional and practical assistance during recovery.

4. Addressing Stigma: Educating others and sharing personal experiences can help combat stigma associated with mpox, fostering a more understanding environment.

5. Mental Health Considerations: Recognizing the emotional impact of living with mpox and seeking professional help when needed is vital for maintaining mental well-being.

6. Building Resilience: Engaging in mindfulness practices and establishing a daily routine can enhance coping strategies and promote a sense of normalcy.

Future Perspectives on Mpox Research

1. Vaccine Development: Ongoing research into effective vaccines will be critical for preventing future outbreaks and controlling the spread of mpox.

2. Treatment Options: Continued exploration of antiviral treatments and therapeutic

interventions will enhance patient outcomes and manage severe cases more effectively.

3. Epidemiological Studies: Comprehensive studies on the transmission dynamics and risk factors associated with mpox will inform public health strategies and interventions.

4. Public Health Policies: Research aimed at developing robust public health policies will be essential to mitigate the impact of mpox outbreaks in communities.

5. Global Collaboration: Strengthening global partnerships among health organizations can facilitate better surveillance, response strategies, and resource sharing in addressing mpox.

6. Education and Awareness: Future initiatives should focus on increasing public awareness and education about mpox to reduce stigma and promote informed health choices.

By focusing on these areas, the future of mpox research holds promise for improved prevention, treatment, and societal understanding, ultimately leading to better health outcomes for affected individuals.

CHAPTER 9

REFERENCES

Suggested Readings and Resources

1. Centers for Disease Control and Prevention (CDC)
    - Website:https://www.cdc.gov/poxvirus/mpox/index.html)
    - Description: Comprehensive information on mpox, including symptoms, transmission, and prevention strategies.

2. World Health Organization (WHO)
    - Website:(https://www.who.int/news-room/fact-sheets/detail/monkeypox)
    - Description: Global health guidelines and updates on mpox outbreaks, vaccination, and treatment options.

3. National Institutes of Health (NIH)
   - Website:(https://www.nih.gov/news-events/news-releases/nih-study-shows-monkeypox-vaccine-effective-preventing-infection)
   - Description: Research findings and clinical studies related to mpox and its vaccine efficacy.

4. American Public Health Association (APHA)
   - Website:(https://www.apha.org/topics-and-issues/infectious-diseases/monkeypox)
   - Description: Public health resources, articles, and advocacy related to infectious diseases, including mpox.

5. Books
   - "Infectious Diseases: A Clinical Short Course" by Frederick S. Southwick
   - Overview: Provides insights into various infectious diseases, including emerging infections like mpox.

6. Support Groups
    - The Trevor Project
    - Website:(https://www.thetrevorproject.org/)
    - Description: Offers support for LGBTQ+ individuals, including those dealing with health-related stigma.

7. Mental Health Resources
    - National Alliance on Mental Illness (NAMI)
    - Website: [NAMI](https://nami.org/)
    - Description: Provides information and support for mental health challenges associated with chronic illness.

8. Scientific Journals
    - Journal of Infectious Diseases
    - Description: Publishes research articles on infectious diseases, including recent findings related to mpox.

9. Online Courses
    - Coursera: Infectious Disease Epidemiology
    - Website:(https://www.coursera.org)
    - Description: Offers courses on the epidemiology of infectious diseases, which can provide deeper insights into diseases like mpox.

10. Podcasts
    - "This Week in Virology"
    - Overview: Discusses current viral outbreaks and research, including episodes focused on emerging viruses like mpox.

These resources offer a wealth of information for individuals seeking to understand mpox better, manage their health, and connect with supportive communities.

CHAPTER 10

APPENDICES

Glossary of Terms

1. Mpox (Monkeypox): A viral disease caused by the monkeypox virus, characterized by fever, rash, and swollen lymph nodes.

2. Transmission: The process through which a disease is spread from one individual to another, which can occur through direct contact, respiratory droplets, or contaminated surfaces.

3. Incubation Period: The time between exposure to the virus and the onset of symptoms, typically ranging from 5 to 21 days for mpox.

4. Vaccination: The administration of a vaccine to stimulate an individual's immune response against a specific pathogen.

5. Endemic: A disease or condition regularly found and consistently present in a particular geographic area.

6. Epidemic: An increase in the number of cases of a disease above what is normally expected in a specific area.

7. Zoonotic Disease: A disease that can be transmitted from animals to humans.

8. Quarantine: The practice of isolating individuals who may have been exposed to a contagious disease to prevent its spread.

9. Public Health: The science of protecting and improving the health of communities through education, policy-making, and research for disease and injury prevention.

10. Contact Tracing: The process of identifying and notifying individuals who may have come into contact with an infected person.

Contact Information for Health Organizations

1. Centers for Disease Control and Prevention (CDC)
    - Website:(https://www.cdc.gov/poxvirus/mpox/index.html)
    - Phone: 1-800-CDC-INFO (1-800-232-4636)

2. World Health Organization (WHO)
    - Website:(https://www.who.int/news-room/fact-sheets/detail/monkeypox)
    - Phone: +41 22 791 2111

3. National Institutes of Health (NIH)
    - Website: [NIH](https://www.nih.gov)
    - Phone: 301-496-4000

4. American Public Health Association (APHA)
    - Website: [APHA](https://www.apha.org)
    - Phone: 202-777-APHA (202-777-2742)

5. National Alliance on Mental Illness (NAMI)
    - Website: [NAMI](https://nami.org)
    - Phone: 1-800-950-NAMI (1-800-950-6264)

6. The Trevor Project
    - Website:(https://www.thetrevorproject.org/)
    - Phone: 1-866-488-7386

7. Local Health Departments
    - To find your local health department, visit [NACCHO](https://www.naccho.org/membership/lhd-directory).

These resources provide essential information and support for individuals seeking guidance on mpox and related health concerns.

THANK YOU SO MUCH READER FOR READING THIS BOOK, PLEASE DROP A REVIEW SO TO ENCOURAGE ME OF THE WORK

www.ingramcontent.com/pod-product-compliance
Lightning Source LLC
Chambersburg PA
CBHW070419230526
45471CB00006B/2888